Florian Ic

COLD
NUCLEAR
FUSION

CREATE SPACE
PUBLISHER

USA 2012

Scientific reviewer:

Dr. Veturia CHIROIU
Honorific member of
Technical Sciences Academy of Romania (ASTR)
PhD supervisor in Mechanical Engineering

Copyright
Title book: Cold Nuclear Fusion
Author book: Florian Ion Petrescu

ISBN 978-1-4782-3426-5

WELCOME

Nuclear fusion is the process by which two or more atomic nuclei join together, or "fuse", to form a single heavier nucleus.

During this process, matter is not conserved because some of the mass of the fusing nuclei is converted to energy which is released.

The binding energy of the resulting nucleus is greater than the binding energy of each of the nuclei that fused to produce it.

Fusion is the process that powers active stars.

Creating the required conditions for fusion on Earth is very difficult, to the point that it has not been accomplished at any scale for protium, the common light isotope of hydrogen that undergoes natural fusion in stars.

In nuclear weapons, some of the energy released by an atomic bomb (fission bomb) is used for compressing and heating a fusion fuel containing heavier isotopes of hydrogen, and also sometimes lithium, to the point of "ignition".

At this point, the energy released in the fusion reactions is enough to briefly maintain the reaction.

Fusion-based nuclear power experiments attempt to create similar conditions using far lesser means, although to date these experiments have failed to maintain conditions needed for ignition long enough for fusion to be a viable commercial power source.

There are many experiments examining the possibility of fusion power for electrical generation.

Nuclear fusion has great potential as a sustainable energy source. This is due to the abundance of hydrogen on the planet and the inert nature of helium (the nucleus which would result from the nuclear fusion of hydrogen atoms).

Unfortunately, a controlled nuclear fusion reaction has not yet been achieved, due to the temperatures required to sustain one.

In hot fusion it need a temperature of 4000 million degrees. Without a minimum of 3000 million degrees we can't make the hot fusion reaction, to obtain the nuclear power. Today we have just 150 million degrees made. To replace the lack of necessary temperature, it uses various tricks.

Because obtaining the necessary huge temperature for hot fusion is still difficult, it is time to focus us on cold nuclear fusion.

We need to bomb the fuel with accelerated deuterium nuclei.

The fuel will be made from heavy water and lithium.

The optimal proportion of lithium will be tested.

It would be preferable to keep fuel in the plasma state.

Research into developing controlled thermonuclear fusion for civil purposes also began in earnest in the 1950s, and it continues to this day. Two projects, the National Ignition Facility and ITER are in the process of reaching breakeven after 60 years of design improvements developed from previous experiments.

The best results were obtained with the Tokamak-type installations

You are welcome to read the full book.

Introduction

Nuclear fusion is the process by which two or more atomic nuclei join together, or "fuse", to form a single heavier nucleus. During this process, matter is not conserved because some of the mass of the fusing nuclei is converted to energy which is released. The binding energy of the resulting nucleus is greater than the binding energy of each of the nuclei that fused to produce it. Fusion is the process that powers active stars.

There are many experiments examining the possibility of fusion power for electrical generation. Nuclear fusion has great potential as a sustainable energy source. This is due to the abundance of hydrogen on the planet and the inert nature of helium (the nucleus which would result from the nuclear fusion of hydrogen atoms). Unfortunately, a controlled nuclear fusion reaction has not yet been achieved, due to the temperatures required to sustain one.

Some fusion techniques can be employed in the design of atomic weaponry and although more generally, it is fission and not fusion, that is associated with the making of the atomic bomb. It is worth noting that fusion can also have a role to play in the design of the hydrogen bomb.

The fusion of two nuclei with lower masses than iron (which, along with nickel, has the largest binding energy per nucleon) generally releases energy, while the fusion of nuclei heavier than iron absorbs energy. The opposite is true for the reverse process, nuclear fission. This means that fusion generally occurs for lighter elements only, and likewise, that fission normally occurs only for heavier elements. There are extreme astrophysical events that can lead to short periods of fusion with heavier nuclei. This is the process that gives rise to nucleosynthesis, the creation of the heavy elements during events such as supernovae.

Creating the required conditions for fusion on Earth is very difficult, to the point that it has not been accomplished at any scale for protium, the common light isotope of hydrogen that undergoes natural fusion in stars. In nuclear weapons, some of the energy released by an atomic bomb (fission bomb) is used for compressing and heating a fusion fuel containing heavier isotopes of hydrogen, and also sometimes lithium, to the point of "ignition". At this point, the energy released in the fusion reactions is enough to briefly maintain the reaction. Fusion-based nuclear power experiments attempt to create similar conditions using far lesser means, although to date these experiments have failed to maintain conditions needed for ignition long enough for fusion to be a viable commercial power source.

Building upon the nuclear transmutation experiments by Ernest Rutherford, carried out several years earlier, the laboratory fusion of heavy hydrogen isotopes was first accomplished by Mark Oliphant in 1932. During the remainder of that decade the steps of the main cycle of nuclear fusion in stars were worked out by Hans Bethe. Research into fusion for military purposes began in the early 1940s as part of the Manhattan Project, but this was not accomplished until 1951 (see the Greenhouse Item nuclear test), and nuclear fusion on a large scale in an explosion was first carried out on November 1, 1952, in the Ivy Mike hydrogen bomb test.

Research into developing controlled thermonuclear fusion for civil purposes also began in earnest in the 1950s, and it continues to this day. Two projects, the National Ignition Facility and ITER are in the process of reaching breakeven after 60 years of design improvements developed from previous experiments.

The best results were obtained with the Tokamak-type installations (see the Figures below).

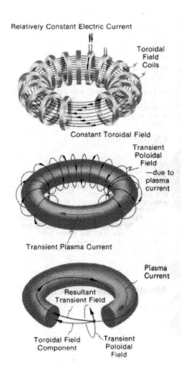

ITER: the world's largest Tokamak

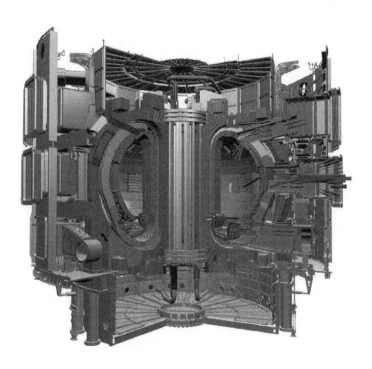

ITER is based on the 'tokamak' concept of magnetic confinement, in which the plasma is contained in a doughnut-shaped vacuum vessel. The fuel—a mixture of deuterium and tritium, two isotopes of hydrogen—is heated to temperatures in excess of 150 million°C, forming a hot plasma. Strong magnetic fields are used to keep the plasma away from the walls; these are produced by superconducting coils surrounding the vessel, and by an electrical current driven through the plasma.

The origin of the energy released in fusion of light elements is due to an interplay of two opposing forces, the nuclear force which draws together protons and neutrons, and the Coulomb force which causes protons to repel each other.

The protons are positively charged and repel each other but they nonetheless stick together, portraying the existence of another force referred to as a nuclear attraction. The strong nuclear force, that overcomes electric repulsion in a very close range.

The effect of this force is not observed outside the nucleus. Hence the force has a strong dependence on distance making it a short range force.

The same force also pulls the neutrons together, or neutrons and protons together. Because the nuclear force is stronger than the Coulomb force for atomic nuclei smaller than iron and nickel, building up these nuclei from lighter nuclei by fusion releases the extra energy from the net attraction of these particles.

For larger nuclei, however, no energy is released, since the nuclear force is short-range and cannot continue to act across still larger atomic nuclei. Thus, energy is no longer released when such nuclei are made by fusion (instead, energy is absorbed in such processes).

Fusion reactions of light elements power the stars and produce virtually all elements in a process called nucleosynthesis. The fusion of lighter elements in stars releases energy (and the mass that always accompanies it).

For example, in the fusion of two hydrogen nuclei to form helium, seven-tenths of 1 percent of the mass is carried away from the system in the form of kinetic energy or other forms of energy (such as electromagnetic radiation). However, the production of elements heavier than iron absorbs energy.

Research into controlled fusion, with the aim of producing fusion power for the production of electricity, has been conducted for over 60 years. It has been accompanied by extreme scientific and technological difficulties, but has resulted in progress.

At present, controlled fusion reactions have been unable to produce break-even (self-sustaining) controlled fusion reactions.

Workable designs for a reactor that theoretically will deliver ten times more fusion energy than the amount needed to heat up plasma to required temperatures (see ITER) were originally scheduled to be operational in 2018, however this has been delayed and a new date has not been stated.

It takes considerable energy to force nuclei to fuse, even those of the lightest element, hydrogen.

This is because all nuclei have a positive charge (due to their protons), and as like charges repel, nuclei strongly resist being put too close together.

Accelerated to high speeds (that is, heated to thermonuclear temperatures), they can overcome this electrostatic repulsion and get close enough for the attractive nuclear force to be sufficiently strong to achieve fusion.

The fusion of lighter nuclei, which creates a heavier nucleus and often a free neutron or proton, generally releases more energy than it takes to force the nuclei together; this is an exothermic process that can produce self-sustaining reactions.

The US National Ignition Facility, which uses laser-driven inertial confinement fusion, is thought to be capable of break-even fusion.

Energy released in most nuclear reactions is much larger than in chemical reactions, because the binding energy that holds a nucleus together is far greater than the energy that holds electrons to a nucleus.

For example, the ionization energy gained by adding an electron to a hydrogen nucleus is 13.6 eV—less than one-millionth of the 17 MeV released in the deuterium–tritium (D–T) reaction shown in the diagram to the right.

Fusion reactions have an energy density many times greater than nuclear fission; the reactions produce far greater energies per unit of mass even though individual fission reactions are generally much more energetic than individual fusion ones, which are themselves millions of times more energetic than chemical reactions.

Only direct conversion of mass into energy, such as that caused by the annihilation collision of matter and antimatter, is more energetic per unit of mass than nuclear fusion.

A substantial energy barrier of electrostatic forces must be overcome before fusion can occur. At large distances two naked nuclei repel one another because of the repulsive electrostatic force between their positively charged protons.

If two nuclei can be brought close enough together, however, the electrostatic repulsion can be overcome by the attractive nuclear force, which is stronger at close distances.

When a nucleon such as a proton or neutron is added to a nucleus, the nuclear force attracts it to other nucleons, but primarily to its immediate neighbours due to the short range of the force.

The nucleons in the interior of a nucleus have more neighboring nucleons than those on the surface.

Since smaller nuclei have a larger surface area-to-volume ratio, the binding energy per nucleon due to the nuclear force generally increases with the size of the nucleus but approaches a limiting value corresponding to that of a nucleus with a diameter of about four nucleons.

It is important to keep in mind that the above picture is a toy model because nucleons are quantum objects, and so, for example, since two neutrons in a nucleus are identical to each other, distinguishing one from the other, such as which one is in the interior and which is on the surface, is in fact meaningless, and the inclusion of quantum mechanics is necessary for proper calculations.

The electrostatic force, on the other hand, is an inverse-square force, so a proton added to a nucleus will feel an electrostatic repulsion from all the other protons in the nucleus.

The electrostatic energy per nucleon due to the electrostatic force thus increases without limit as nuclei get larger.

H-hour

With the help of powerful lasers one can create a dense and highly ionized plasma. We need a highly ionized dense plasma to achieve nuclear fusion (cold or hot).

Since 1989, it talks about achieving nuclear fusion hot and cold. Another two decades have passed and humanity still does not benefit from nuclear fusion energy.

What actually happens? Is it an unattainable myth? It was also circulated by the media that has been achieved nuclear fusion heat. Since 1989 there are all sorts of scientists with all kinds of crafted devices, which declare that they can produce nuclear power obtained by cold fusion (using cold plasma).

May be that these devices works, but their yield is probably too small, or at an enlarged scale these give not the expected results. This is the real reason why we can't use yet the survival fuel (the deuterium).

Unfortunately today the dominant processes that produce energy are combustion (reaction) chemical combination of carbon with oxygen. Thermal energy released from such reactions is conventionally valued at about 7000 calories per gram.

Only the early 20th century physicists have succeeded in producing, other energy than by traditional methods. Energy release per unit mass was enormous compared with that obtained by conventional procedures.

The Kilowatt based on nuclear fission of uranium nuclei has today a significant share in global energy balance.

Unfortunately, the nuclear power plants burn the fuel uranium, already considered conventional and on extinct.

The current nuclear power is considered a transition way, to the energy thermonuclear, based on fusion of light nuclei.

The main particularity of synthesis reaction (fusion) is the high prevalence of the used fuel (primary), deuterium. It can be obtained relatively simply from ordinary water.

Deuterium was extracted from water for the first time by Harold Urey in 1931. Even at that time, small linear electrostatic accelerators, have indicated that D-D reaction (fusion of two deuterium nuclei) is exothermic.

Today we know that not only the first isotope of hydrogen (deuterium) produces fusion energy, but and the second (heavy) isotope of hydrogen (tritium) can produce energy by nuclear fusion.

The first reaction is possible between two nuclei of deuterium, from which can be obtained, either a tritium nucleus plus a proton and energy, or an isotope of helium with a neutron and energy.

$$
{}_{1}^{2}D + {}_{1}^{2}D \rightarrow
\begin{cases}
{}_{1}^{3}T + 1MeV + {}_{1}^{1}H + 3MeV = {}_{1}^{3}T + {}_{1}^{1}H + 4MeV & (1) \\
{}_{2}^{3}He + 0.8MeV + {}^{1}n + 2.5MeV = {}_{2}^{3}He + {}^{1}n + 3.3MeV & (2)
\end{cases}
$$

Observations: a deuterium nucleus has a proton and a neutron; a tritium nucleus has a proton and two neutrons.

Fusion can occur between a nucleus of deuterium and one of tritium.

$$_1^2 D +_1^3 T \rightarrow_2^4 He + 3.5 MeV +_1^1 n + 14 MeV =_2^4 He+_1^1 n + 17.5 MeV \quad (3)$$

Another fusion reaction can be produced between a nucleus of deuterium and an isotope of helium.

$$_1^2 D +_2^3 He \rightarrow_2^4 He + 3.7 MeV +_1^1 H + 14.7 MeV =_2^4 He+_1^1 H + 18.4 MeV \quad (4)$$

For these reactions to occur, should that the deuterium nuclei have enough kinetic energy to overcome the electrostatic forces of rejection due to the positive tasks of protons in the nuclei.

For deuterium, for average kinetic energy are required tens of keV.

For 1 keV are needed about 10 million degrees temperature. For this reason hot fusion requires a temperature of hundreds of millions of degrees.
The huge temperature is done with high power lasers acting hot plasma.
Electromagnetic fields are arranged so that it can maintain hot plasma.

The best results were obtained with the Tokamak-type installations.
ITER: the world's largest Tokamak

ITER is based on the 'tokamak' concept of magnetic confinement, in which the plasma is contained in a doughnut-shaped vacuum vessel. The fuel—a mixture of deuterium and tritium, two isotopes of hydrogen—is heated to temperatures in excess of 150 million°C, forming a hot plasma. Strong magnetic fields are used to keep the plasma away from the walls; these are produced by superconducting coils surrounding the vessel, and by an electrical current driven through the plasma.

Deuterium fuel is delivered in heavy water, D_2O.
Tritium is obtained in the laboratory by the following reaction.

$$_3^6 Li +\,^1 n \rightarrow\,_1^3 T +\,_2^4 He + 4.6 MeV \quad (5)$$

Lithium, the third element in Mendeleev's table, is found in nature in sufficient quantities.
The accelerated neutrons which produce the last presented reaction with lithium, appear from the second and the third presented reaction.
Raw materials for fusion are deuterium and lithium.

All fusion reactions shown produce finally energy and He. He is a (gas) inert element. Because of this, fusion reaction is clean, and far superior to nuclear fission.

Hot fusion works with very high temperatures.

In cold fusion, it must accelerate the deuterium nucleus, in linear or circular accelerators. Final energy of accelerated deuterium nuclei should be well calibrated for a positive final yield of fusion reactions (more mergers, than fission). Electromagnetic fields which maintain the plasma (cold and especially the warm), should be and constrictors (especially at cold fusion), for to press, and more close together the nuclei.

The potential energy with that two particles reject each other, can be approximately calculated with the following relationship.

$$U \equiv E_p = \frac{1}{4 \cdot \pi \cdot \varepsilon_0} \cdot \frac{q_1 \cdot q_2}{d_{12}} = \frac{1}{4 \cdot \pi \cdot 8.8541853 \cdot 10^{-12}} \cdot \frac{\left(1.602 \cdot 10^{-19}\right)^2}{4 \cdot 10^{-15}} =$$

$$= 5.7664 \cdot 10^{-14} [J] = 5.7664 \cdot 10^{-14} \cdot 6.242 \cdot 10^{18} [eV] = 3.599 \cdot 10^5 [eV] =$$

$$= 360 [keV] \qquad (6)$$

At a keV is necessary a temperature of 10 million ^{0}C.
At 360 keV is necessary a temperature of 3600 million ^{0}C.

In hot fusion it need a temperature of 3600 million degrees.
Without a minimum of 3000 million degrees we can't make the hot fusion reaction, to obtain the nuclear power.

Today we have just 150 million degrees made.

To replace the lack of necessary temperature, it uses various tricks.

In cold fusion one must accelerate the deuterium nuclei at an energy of 360 [keV], and then collide them with the cold fusion fuel (heavy water and lithium).

Cold Nuclear Fusion

Because obtaining the necessary huge temperature for hot fusion is still difficult, it is time to focus us on cold nuclear fusion.
We need to bomb the fuel with accelerated deuterium nuclei.
The fuel will be made from heavy water and lithium.
The optimal proportion of lithium will be tested.
It would be preferable to keep fuel in the plasma state.
Between deuterium and tritium the smallest radius is the radius of deuterium nucleus.

Deuterium $\quad A = 2 \quad A^{1/3} = 1.259921 \Rightarrow R_D = 1.8268855223476 \cdot 10^{-15}[m]$

Tritium $\quad A = 3 \quad A^{1/3} = 1.44224957 \Rightarrow R_T = 2.0912618769457 \cdot 10^{-15}[m]$

We calculate the minimum distance between two particles which meet together.
This is just the diameter of a deuterium nucleus, d_{12D}.

$$d_{12D} = 2 \cdot R_D = 2 \cdot 1.8268855223476 \cdot 10^{-15}[m] =$$
$$= 3.6537710446952 \cdot 10^{-15}[m] =$$
$$\approx 3.653771 \cdot 10^{-15}[m]$$

The deuterium nuclei which will bomb the nuclear fuel, will be accelerated with the (least) energy which reject the two neighboring deuterium nuclei (see the below relationship).

$$U \equiv E_p = \frac{1}{4 \cdot \pi \cdot \varepsilon_0} \cdot \frac{q_1 \cdot q_2}{d_{12D}} = \frac{1}{4 \cdot \pi \cdot 8.8541853 \cdot 10^{-12}} \cdot \frac{\left(1.602 \cdot 10^{-19}\right)^2}{3.653771 \cdot 10^{-15}} =$$

$$= 6.3128464855 \cdot 10^{-14}[J] = 6.3128464855 \cdot 10^{-14} \cdot 6.242 \cdot 10^{18}[eV] =$$

$$= 3.94 \cdot 10^5 [eV] = 3.94 \cdot 10^2 [keV] = 394[keV]$$

Much success!
The Author

Bibliography

^ "Progress in Fusion". ITER. Retrieved 2010-02-15.
^ "The National Ignition Facility: Ushering in a New Age for Science". National Ignition Facility. Retrieved 2009-09-13.
^ "DOE looks again at inertial fusion as potential clean-energy source", David Kramer, Physics Today, March 2011, p 26
^ The Most Tightly Bound Nuclei. Hyperphysics.phy-astr.gsu.edu. Retrieved on 2011-08-17.
^ F. Winterberg "Conjectured Metastable Super-Explosives formed under High Pressure for Thermonuclear Ignition"
^ Zhang, Fan; Murray, Stephen Burke; Higgins, Andrew (2005) "Super compressed detonation method and device to effect such detonation[dead link]"
^ I.I. Glass and J.C. Poinssot "IMPLOSION DRIVEN SHOCK TUBE". NASA
^ D.Sagie and I.I. Glass (1982) "Explosive-driven hemispherical implosions for generating fusion plasmas"
^ T. Saito, A. K. Kudian and I. I. Glass "Temperature Measurements Of An Implosion Focus"
^ S.E. Jones (1986). "Muon-Catalysed Fusion Revisited". Nature 321 (6066): 127–133. Bibcode 1986Natur.321..127J. DOI:10.1038/321127a0.
^ Access: Desktop fusion is back on the table: Nature News. Nature.com. Retrieved on 2011-08-17.
^ Supplementary methods for "Observation of nuclear fusion driven by a pyroelectric crystal". Main article Naranjo, B.; Gimzewski, J.K.; Putterman, S. (2005). "Observation of nuclear fusion driven by a

pyroelectric crystal". Nature 434 (7037): 1115–1117. Bibcode 2005Natur.434.1115N. DOI:10.1038/nature03575. PMID 15858570.
^ UCLA Crystal Fusion. Rodan.physics.ucla.edu. Retrieved on 2011-08-17.
^ Phil Schewe and Ben Stein (2005). "Pyrofusion: A Room-Temperature, Palm-Sized Nuclear

Important reactions

The most important fusion process in nature is the one that powers stars.

The net result is the fusion of four protons into one alpha particle, with the release of two positrons, two neutrinos (which changes two of the protons into neutrons), and energy, but several individual reactions are involved, depending on the mass of the star.

For stars the size of the sun or smaller, the proton-proton chain dominates. In heavier stars, the CNO cycle is more important.

Both types of processes are responsible for the creation of new elements as part of stellar nucleosynthesis (see the Figures below).

At the temperatures and densities in stellar cores the rates of fusion reactions are notoriously slow.

For example, at solar core temperature ($T \approx$ 15 MK) and density (160 g/cm3), the energy release rate is only 276 µW/cm3—about a quarter of the volumetric rate at which a resting human body generates heat.

Thus, reproduction of stellar core conditions in a lab for nuclear fusion power production is completely impractical.

Because nuclear reaction rates strongly depend on temperature ($\exp(-E/kT)$), achieving reasonable energy production rates in terrestrial fusion reactors requires 10–100 times higher temperatures (compared to stellar interiors): $T \approx 0.1$–1.0 GK.

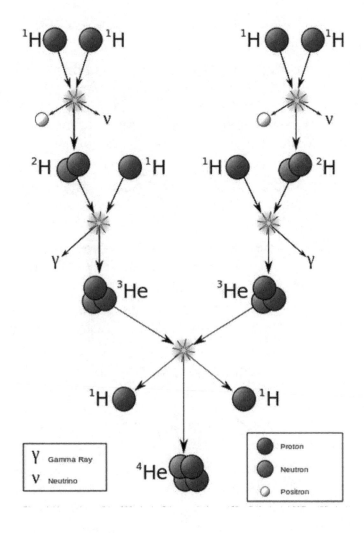

The proton-proton chain dominates in stars the size
of the Sun or smaller

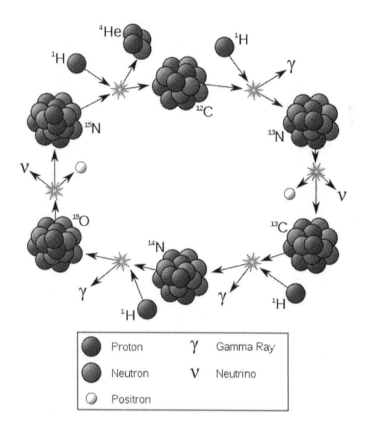

The CNO cycle dominates in stars heavier than the Sun

In man-made fusion, the primary fuel is not constrained to be protons and higher temperatures can be used, so reactions with larger cross-sections are chosen.

This implies a lower Lawson criterion, and therefore less startup effort. Another concern is the production of neutrons, which activate the reactor structure radiologically, but also have the advantages of allowing volumetric extraction of the fusion energy and tritium breeding.

Reactions that release no neutrons are referred to as aneutronic.

To be a useful energy source, a fusion reaction must satisfy several criteria. It must

Be exothermic: This may be obvious, but it limits the reactants to the low Z (number of protons) side of the curve of binding energy. It also makes helium 4He the most common product because of its extraordinarily tight binding, although 3He and 3H also show up.

Involve low Z nuclei: This is because the electrostatic repulsion must be overcome before the nuclei are close enough to fuse.

Have two reactants: At anything less than stellar densities, three body collisions are too improbable. In inertial confinement, both stellar densities and temperatures are exceeded to compensate for the shortcomings of the third parameter of the Lawson criterion, ICF's very short confinement time.

Have two or more products: This allows simultaneous conservation of energy and momentum without relying on the electromagnetic force.

Conserve both protons and neutrons: The cross sections for the weak interaction are too small.

Few reactions meet these criteria. The following are those with the largest cross sections.

(1) $^2_1D + ^3_1T \to ^4_2He$ (3.5 MeV) + n^0 (14.1 MeV)

(2i) $^2_1D + ^2_1D \to ^3_1T$ (1.01 MeV) + p^+ (3.02 MeV)   50%

(2ii) $\to ^3_2He$ (0.82 MeV) + n^0 (2.45 MeV)   50%

(3) $^2_1D + ^3_2He \to ^4_2He$ (3.6 MeV) + p^+ (14.7 MeV)

(4) $^3_1T + ^3_1T \to ^4_2He$ + 2 n^0 + 11.3 MeV

(5) $^3_2He + ^3_2He \to ^4_2He$ + 2 p^+ + 12.9 MeV

(6i) $^3_2He + ^3_1T \to ^4_2He$ + p^+ + n^0 + 12.1 MeV   51%

(6ii) $\to ^4_2He$ (4.8 MeV) + ^2_1D (9.5 MeV)   43%

(6iii) $\to ^4_2He$ (0.5 MeV) + n^0 (1.9 MeV) + p^+ (11.9 MeV)   6%

(7i) $^2_1D + ^6_3Li \to 2\,^4_2He + 22.4$ MeV

(7ii) $\to ^3_2He + ^4_2He$ + n^0 + 2.56 MeV

(7iii) $\to ^7_3Li + p^+$ + 5.0 MeV

(7iv) $\to ^7_4Be + n^0$ + 3.4 MeV

(8) $p^+ + ^6_3Li \to ^4_2He$ (1.7 MeV) + ^3_2He (2.3 MeV)

(9) $^3_2He + ^6_3Li \to 2\,^4_2He + p^+$ + 16.9 MeV

(10) $p^+ + ^{11}_5B \to 3\,^4_2He$ + 8.7 MeV

For reactions with two products, the energy is divided between them in inverse proportion to their masses, as shown. In most reactions with three products, the distribution of energy varies.

For reactions that can result in more than one set of products, the branching ratios are given.

Some reaction candidates can be eliminated at once. The D-^{6}Li reaction has no advantage compared to p^+-$^{11}_5$B because it is roughly as difficult to burn but produces substantially more neutrons through $^2_1$D-$^2_1$D side reactions.

There is also a p^+-$^7_3$Li reaction, but the cross section is far too low, except possibly when Ti>1 MeV, but at such high temperatures an endothermic, direct neutron-producing reaction also becomes very significant.

Finally there is also a p^+-$^9_4$Be reaction, which is not only difficult to burn, but $^9_4$Be can be easily induced to split into two alpha particles and a neutron.

In addition to the fusion reactions, the following reactions with neutrons are important in order to "breed" tritium in "dry" fusion bombs and some proposed fusion reactors:

$$n^0 + {}^6_3\text{Li} \rightarrow {}^3_1\text{T} + {}^4_2\text{He}$$
$$n^0 + {}^7_3\text{Li} \rightarrow {}^3_1\text{T} + {}^4_2\text{He} + n^0$$

The actual ratios of fusion to Bremsstrahlung power will likely be significantly lower for several reasons.

For one, the calculation assumes that the energy of the fusion products is transmitted completely to the fuel ions, which then lose energy to the electrons by collisions, which in turn lose energy by Bremsstrahlung.

However, because the fusion products move much faster than the fuel ions, they will give up a significant fraction of their energy directly to the electrons.

Secondly, the ions in the plasma are assumed to be purely fuel ions. In practice, there will be a significant proportion of impurity ions, which will then lower the ratio. In particular, the fusion products themselves must remain in the plasma until they have given up their energy, and will remain some time after that in any proposed confinement scheme.

Finally, all channels of energy loss other than Bremsstrahlung have been neglected. The last two factors are related.

On theoretical and experimental grounds, particle and energy confinement seem to be closely related.

In a confinement scheme that does a good job of retaining energy, fusion products will build up. If the fusion products are efficiently ejected, then energy confinement will be poor, too.

Aneutronic fusion

Aneutronic fusion is any form of fusion power where neutrons carry no more than 1% of the total released energy.

The most-studied fusion reactions release up to 80% of their energy in neutrons.

Successful aneutronic fusion would greatly reduce problems associated with neutron radiation such as ionizing damage, neutron activation, and requirements for biological shielding, remote handling, and safety.

Some proponents also see a potential for dramatic cost reductions by converting energy directly to electricity.

However, the conditions required to harness aneutronic fusion are much more extreme than those required for the conventional deuterium–tritium (DT) fuel cycle.

Candidate aneutronic reactions

There are a few fusion reactions that have no neutrons as products on any of their branches. Those with the largest cross sections are these:

$$D + {}^3He \rightarrow {}^4He\ (3.6\ MeV) + p\ (14.7\ MeV)$$
$$D + {}^6Li \rightarrow 2\ {}^4He + 22.4\ MeV$$
$$p + {}^6Li \rightarrow {}^4He\ (1.7\ MeV) + {}^3He\ (2.3\ MeV)$$
$$^3He + {}^6Li \rightarrow 2\ {}^4He + p + 16.9\ MeV$$
$$^3He + {}^3He \rightarrow {}^4He + 2\ p + 12.86\ MeV$$
$$p + {}^7Li \rightarrow 2\ {}^4He + 17.2\ MeV$$
$$p + {}^{11}B \rightarrow 3\ {}^4He + 8.7\ MeV$$
$$p + {}^{15}N \rightarrow {}^{12}C + {}^4He + 5.0\ MeV$$

The two of these which use deuterium as a fuel produce some neutrons with D–D side reactions.

Although these can be minimized by running hot and deuterium-lean, the fraction of energy released as neutrons will probably be several percent, so that these fuel cycles, although neutron-poor, do not qualify as aneutronic according to the 1% threshold.

The next two reactions' rates (involving p, ^{3}He, and ^{6}Li) are not particularly high in a thermal plasma.

When treated as a chain, however, they offer the possibility of enhanced reactivity due to a non-thermal distribution.

The product ^{3}He from the first reaction could participate in the second reaction before thermalizing, and the product p from the second reaction could participate in the first reaction before thermalizing.

Unfortunately, detailed analyses do not show sufficient reactivity enhancement to overcome the inherently low cross section.

The pure ^{3}He reaction suffers from a fuel-availability problem. ^{3}He occurs in only minuscule amounts naturally on Earth, so it would either have to be bred from neutron reactions (counteracting the potential advantage of aneutronic fusion), or mined from extraterrestrial sources.

The top several meters of the surface of the Moon is relatively rich in ^{3}He, on the order of 0.01 parts per million by weight, but mining this resource and returning it to Earth would be very difficult and expensive. ^{3}He could in principle be recovered from the atmospheres of the gas giant planets, Jupiter, Saturn, Neptune and Uranus, but this would be even more challenging.

The p $-^7$Li reaction has no advantage over p–^{11}B, given its somewhat lower cross section.

For the above reasons, most studies of aneutronic fusion concentrate on the reaction, p – ^{11}B.

Despite the suggested advantages of aneutronic fusion, the vast majority of fusion research has gone toward D-T fusion because the technical challenges of hydrogen–boron (p $-^{11}$B) fusion are so formidable.

Hydrogen–boron fusion requires ion energies or temperatures almost ten times higher than those for D-T fusion. For any given densities of the reacting nuclei, the reaction rate for hydrogen-boron achieves its peak rate at around 600 keV (6.6 billion degrees Celsius or 6.6 gigakelvins).

In addition, the peak reaction rate of p–^{11}B is only one third that for D–T, requiring better plasma confinement.

Confinement is usually characterized by the time τ the energy must be retained so that the fusion power released exceeds the power required to heat the plasma.

Since the confinement properties of conventional fusion approaches, such as the tokamak and laser pellet fusion are marginal, most aneutronic proposals use radically different confinement concepts.

In every published fusion power plant design, the part of the plant that produces the fusion reactions is much more expensive than the part that converts the nuclear power to electricity. In that case, as indeed in most power systems, power density is a very important characteristic. Doubling power density at least halves the cost of electricity. In addition, the confinement time required depends on the power density.

Lawrenceville Plasma Physics has published initial results and outlined a theory and experimental program for aneutronic fusion with the Dense Plasma Focus (DPF), building on earlier discussions. The private effort was initially funded by NASA's Jet Propulsion Laboratory. Support for other DPF aneutronic fusion investigations has come from the Air Force Research Laboratory.

Polywell fusion was pioneered by Robert W. Bussard and funded by the US Navy, uses inertial electrostatic confinement. Research continues at the company he founded, EMC2.

The Z-machine at Sandia National Laboratory, a z-pinch device, can produce ion energies of interest to hydrogen–boron reactions, up to 300 keV. Non-equilibrium plasmas usually have an electron temperature higher than their ion temperature, but the plasma in the Z machine has a special, reverted non-equilibrium state, where ion temperature is 100 times higher than electron temperature. These data represent a new research field, and indicate that Bremsstrahlung losses could be in fact lower than previously expected in such a design.

None of these efforts has yet tested its device with hydrogen–boron fuel, so the anticipated performance is based on extrapolating from theory, experimental results with other fuels and from simulations.

A picosond laser produced hydrogen–boron aneutronic fusions for a Russian team in 2005. However, the number of the resulting α particles (around 103 per laser pulse) was extremely low.

Aneutronic fusion reactions produce the overwhelming bulk of their energy in the form of charged particles instead of neutrons.

This means that energy could be converted directly into electricity by various techniques.

Many proposed direct conversion techniques are based on mature technology derived from other fields, such as microwave technology, and some involve equipment that is more compact and potentially cheaper than that involved in conventional thermal production of electricity.

In contrast, fusion fuels like deuterium-tritium (DT), which produce most of their energy in the form of neutrons, require a standard thermal cycle, in which the neutrons are used to boil water, and the resulting steam drives a large turbine and generator.

This equipment is sufficiently expensive that about 80% of the capital cost of a typical fossil-fuel electric power generating station is in the thermal conversion equipment.

Thus, fusion with DT fuels could not significantly reduce the capital costs of electric power generation even if the fusion reactor that produces the neutrons were cost-free. (Fuel costs would, however, be greatly reduced.) But according to proponents, aneutronic fusion with direct electric conversion could, in theory, produce electricity with reduced capital costs.

Direct conversion techniques can either be inductive, based on changes in magnetic fields, or electrostatic, based on making charged particles work against an electric field.

If the fusion reactor worked in a pulsed mode, inductive techniques could be used.

A sizable fraction of the energy released by aneutronic fusion would not remain in the charged fusion products but would instead be radiated as X-rays.

Some of this energy could also be converted directly to electricity.

Because of the photoelectric effect, X-rays passing through an array of conducting foils would transfer some of their energy to electrons, which can then be captured electrostatically.

Since X-rays can go through far greater thickness of material than electrons can, many hundreds or even thousands of layers would be needed to absorb most of the X-rays.

Dense plasma focus

A dense plasma focus (DPF) is a machine that produces, by electromagnetic acceleration and compression, a short-lived plasma that is so hot and dense that it can cause nuclear fusion and emit X-rays.

The electromagnetic compression of the plasma is called a pinch.

It was invented in the early 1960s by J.W. Mather and also independently by N.V. Filippov.

The plasma focus is similar to the high-intensity plasma gun device (HIPGD) (or just plasma gun), which ejects plasma in the form of a plasmoid, without pinching it.

Intense bursts of X-rays and charged particles are emitted, as are nuclear fusion neutrons, when operated using deuterium.

There is ongoing research that demonstrates potential applications as a soft X-ray source for next-generation microelectronics lithography, surface micromachining, pulsed X-ray and neutron source for medical and security inspection applications and materials modification, among others.

For nuclear weapons applications, dense plasma focus devices can be used as an external neutron source.

Other applications include simulation of nuclear explosions (for testing of the electronic equipment) and a short and intense neutron source useful for non-contact discovery or inspection of nuclear materials (uranium, plutonium).

An important characteristic of the dense plasma focus is that the energy density of the focused plasma is practically a constant over the whole range of machines, from sub-kilojoule machines to megajoule machines, when these machines are tuned for optimal operation.

This means that a small table-top-sized plasma focus machine produces essentially the same plasma characteristics (temperature and density) as the largest plasma focus. Of course the larger machine will produce the larger volume of focused plasma with a corresponding longer lifetime and more radiation yield.

Even the smallest plasma focus has essentially the same dynamic characteristics as larger machines, producing the same plasma characteristics and the same radiation products. This is due to the scalability of plasma phenomena.

See also plasmoid, the self-contained magnetic plasma ball that may be produced by a dense plasma focus.

The charged bank of electrical capacitors (also called a Marx bank or Marx generator) is switched onto the anode.

The gas breaks down. A rapidly rising electric current flows across the backwall electrical insulator, axisymmetrically, as depicted by the path (labeled 1).

The axisymmetric sheath of plasma current lifts off the insulator due to the interaction of the current with its own magnetic field (Lorentz force).

The plasma sheath is accelerated axially, to position 2, and then to position 3, ending the axial phase of the device.

The whole process proceeds at many times the speed of sound in the ambient gas.

As the current sheath continues to move axially, the portion in contact with the anode slides across the face of the anode, axisymmetrically.

When the imploding front of the shock wave coalesces onto the axis, a reflected shock front emanates from the axis until it meets the driving current sheath which then forms the axisymmetric boundary of the pinched, or focused, hot plasma column.

The dense plasma column (akin to the Z-pinch) rapidly pinches and undergoes instabilities and breaks up.

The intense electromagnetic and particle bursts, collectively referred to as multi-radiation occur during the dense plasma and breakup phases.

These critical phases last typically tens of nanoseconds for a small (kJ, 100 kA) focus to around a microsecond for a large (MJ, several MA) focus.

The whole process, including axial and radial phases, may last, for the Mather DPF, a few microseconds (for a small focus) to 10 microseconds (for a large focus).

A Filippov focus has a very short axial phase compared to a Mather focus.

A network of ten identical DPF machines operates in eight countries around the world.

This network produces research papers on topics including machine optimization & diagnostics (soft x-rays, neutrons, electron and ion beams), applications (microlithography, micromachining, materials modification and fabrication, imaging & medical, astrophysical simulation) as well as modeling & computation.

The network was organized by Sing Lee in 1986 and is coordinated by the Asian African Association for Plasma Training, AAAPT.

A simulation package, the Lee Model, has been developed for this network but is applicable to all plasma focus devices.

The code typically produces excellent agreement between computed and measured results, and is available for downloading as a Universal Plasma Focus Laboratory Facility.

The Institute for Plasma Focus Studies IPFS was founded on 25 February 2008 to promote correct and innovative use of the Lee Model code and to encourage the application of plasma focus numerical experiments. IPFS research has already extended numerically-derived neutron scaling laws to multi-megajoule experiments. These await verification.

Numerical experiments with the code have also resulted in the compilation of a global scaling law indicating that the well-known neutron saturation effect is better correlated to a scaling deterioration mechanism.

This is due to the increasing dominance of the axial phase dynamic resistance as capacitor bank impedance decreases with increasing bank energy (capacitance).

In principle, the resistive saturation could be overcome by operating the pulse power system at a higher voltage.

In Argentina there is an Inter-institutional Program for Plasma Focus Research since 1996, coordinated by a National Laboratory of Dense Magnetized Plasmas in Tandil, Buenos Aires.

The Program also cooperates with the Chilean Nuclear Energy Commission, and networks the Argentine National Energy Commission, the Scientific Council of Buenos Aires, the University of Center, the University of Mar del Plata, The University of Rosario, and the Institute of Plasma Physics of the University of Buenos Aires.

The program operates six Plasma Focus Devices, developing applications, in particular ultra-short tomography and substance detection by neutron pulsed interrogation.

Chile currently operates the facility SPEED-2, the largest Plasma Focus facility of the southern hemisphere. PLADEMA also contributed during the last decade with several mathematical models of Plasma Focus.

The thermodynamic model was able to develop for the first time design maps combining geometrical and operational parameters, showing that there is always an optimum gun length and charging pressure which maximize the neutron emission.

Currently there is a complete finite-elements code validated against numerous experiments, which can be used confidently as a design tool for Plasma Focus.

Since the beginning of 2009, a number of new plasma focus machines have been/are being commissioned including the INTI Plasma Focus in Malaysia, the NX3 in Singapore and the first plasma focus to be commissioned in a US university in recent times, the KSU Plasma Focus at Kansas State University which recorded its first fusion neutron emitting pinch on New Year's Eve 2009.

Several groups have proposed that fusion power based on the DPF could be viable, possibly even with low-neutron fuel cycles like p-B11.

The feasibility of net power from p-B11 in the DPF requires that the bremsstrahlung losses be reduced by quantum mechanical effects induced by the powerful magnetic field.

The high magnetic field will also result in a high rate of emission of cyclotron radiation, but at the densities envisioned, where the plasma frequency is larger than the cyclotron frequency, most of this power will be reabsorbed before being lost from the plasma.

Another advantage claimed is the capability of direct conversion of the energy of the fusion products into electricity, with an efficiency potentially above 70%. Experiments and computer simulations to investigate the capability of DPF for fusion power are underway at Lawrenceville Plasma Physics (LPP) under the direction of Eric Lerner, who explained his "Focus Fusion" approach in a 2007 Google Tech Talk.

Fusion power

Fusion power is the power generated by nuclear fusion processes.

In fusion reactions two light atomic nuclei fuse together to form a heavier nucleus (in contrast with fission power).

In doing so they release a comparatively large amount of energy arising from the binding energy due to the strong nuclear force which is manifested as an increase in temperature of the reactants.

Fusion power is a primary area of research in plasma physics.

The term is commonly used to refer to potential commercial production of net usable power from a fusion source, similar to the usage of the term "steam power."

The leading designs for controlled fusion research use magnetic (tokamak design) or inertial (laser) confinement of a plasma, with heat from the fusion reactions used to operate a steam turbine which in turn drives electrical generators, similar to the process used in fossil fuel and nuclear fission power stations.

As of July 2010, the largest experiment by means of magnetic confinement has been the Joint European Torus (JET).

In 1997, JET produced a peak of 16.1 megawatts (21,600 hp) of fusion power (65% of input power), with fusion power of over 10 MW (13,000 hp) sustained for over 0.5 sec.

Its successor, ITER, was officially announced as part of a seven-country consortium. ITER is designed to produce ten times more fusion power than the power put into the plasma. ITER is currently under construction in Cadarache, France.

Inertial (laser) confinement, which was for a time seen as more difficult or infeasible, has generally seen less development effort than magnetic approaches.

However, this approach made a comeback following further innovations, and is being developed at both the United States National Ignition Facility as well as the planned European Union High Power laser Energy Research (HiPER) facility.

NIF reached initial operational status in 2010 and has been in the process of increasing the power and energy of its "shots". Fusion ignition tests are to follow.

Fusion powered electricity generation was initially believed to be readily achievable, as fission power had been.

However, the extreme requirements for continuous reactions and plasma containment led to projections being extended by several decades.

In 2010, more than 60 years after the first attempts, commercial power production is still believed to be unlikely before 2050.

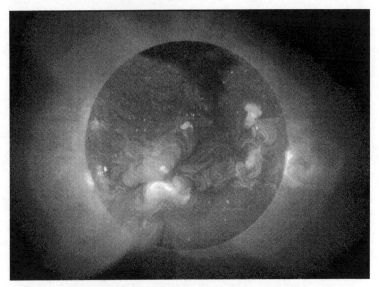

The Sun is a natural fusion reactor

The basic concept behind any fusion reaction is to bring two or more nuclei close enough together so that the residual strong force (nuclear force) in their nuclei will pull them together into one larger nucleus.

If two light nuclei fuse, they will generally form a single nucleus with a slightly smaller mass than the sum of their original masses (though this is not always the case).

The difference in mass is released as energy according to Albert Einstein's mass-energy equivalence formula $E=mc^2$.

If the input nuclei are sufficiently massive, the resulting fusion product will be heavier than the sum of the reactants' original masses, in which case the reaction requires an external source of energy.

The dividing line between "light" and "heavy" is iron-56. Above this atomic mass, energy will generally be released by nuclear fission reactions; below it, by fusion.

Fusion between the nuclei is opposed by the repulsive positive electrical charge common to all nuclei due to protons in the nucleus.

To overcome this electrostatic force, or "Coulomb barrier", the kinetic energy of the atoms must be increased.

The easiest way to do this is to heat the atoms, which has the side effect of stripping the electrons from the atoms and leaving them as bare nuclei.

In most experiments the nuclei and electrons are left in a fluid known as a plasma.

The temperatures required to provide the nuclei with enough energy to overcome their repulsion is a function of the total charge, so hydrogen, which has the smallest nuclear charge therefore fuses at the lowest temperature.

Helium has an extremely low mass per nucleon and therefore is energetically favoured as a fusion product.

As a consequence, most fusion reactions combine isotopes of hydrogen ("protium", deuterium, or tritium) to form isotopes of helium (^{3}He or ^{4}He) as the fusion end product.

D-T fuel cycle

According to the Lawson criterion, the easiest and most immediately promising nuclear reaction for fusion power is:

$$^2_1D + {}^3_1T \rightarrow {}^4_2He + {}^1_0n$$

Hydrogen-2 (Deuterium) is a naturally occurring isotope of hydrogen and is commonly available.

The large mass ratio of the hydrogen isotopes makes their separation easy compared to the difficult uranium enrichment process.

Hydrogen-3 (Tritium) is also an isotope of hydrogen, but it occurs naturally in only negligible amounts due to its half-life of 12.32 years.

Consequently, the deuterium-tritium fuel cycle requires the breeding of tritium from lithium using one of the following reactions:

$$^1_0n + {}^6_3Li \rightarrow {}^3_1T + {}^4_2He$$

$$^1_0n + {}^7_3Li \rightarrow {}^3_1T + {}^4_2He + {}^1_0n$$

The reactant neutron is supplied by the D-T fusion reaction shown above, and the one that has the greatest yield of energy.

The reaction with ^6Li is exothermic, providing a small energy gain for the reactor.

The reaction with ^7Li is endothermic but does not consume the neutron. At least some ^7Li reactions are required to replace the neutrons lost to absorption by other elements.

Most reactor designs use the naturally occurring mix of lithium isotopes.

However, the supply of lithium is relatively limited as other applications such as Li-ion batteries have increased its demand.

Several drawbacks are commonly attributed to D-T fusion power:

It produces substantial amounts of neutrons that result in the neutron activation of the reactor materials.

Only about 20% of the fusion energy yield appears in the form of charged particles with the remainder carried off by neutrons, which limits the extent to which direct energy conversion techniques might be applied.

The use of D-T fusion power depends on lithium resources, which are less abundant than deuterium resources. However, lithium is relatively abundant on earth.

It requires the handling of the radioisotope tritium. Similar to hydrogen, tritium is difficult to contain and may leak from reactors in some quantity.

Some estimates suggest that this would represent a fairly large environmental release of radioactivity.

The neutron flux expected in a commercial D-T fusion reactor is about 100 times that of current fission power reactors, posing problems for material design.

Design of suitable materials is under way but their actual use in a reactor is not proposed until the design generation after ITER.

After a single series of D-T tests at JET, the largest fusion reactor yet to use this fuel, the vacuum vessel was sufficiently radioactive that remote handling was required for the year following the tests.

In a production setting, the neutrons would be used to react with lithium in order to create more tritium.

This also deposits the energy of the neutrons in the lithium, which would then be transferred to drive electrical production.

The lithium neutron absorption reaction protects the outer portions of the reactor from the neutron flux.

Newer designs, the advanced tokamak in particular, also use lithium inside the reactor core as a key element of the design. The plasma interacts directly with the lithium, preventing a problem known as "recycling". The advantage of this design was demonstrated in the Lithium Tokamak Experiment.

D-D fuel cycle

Though more difficult to facilitate than the deuterium-tritium reaction, fusion can also be achieved through the reaction of deuterium with itself.

This reaction has two branches that occur with nearly equal probability:

$$^2_1D + {}^2_1D \rightarrow {}^3_1T + {}^1_1H$$

$$^2_1D + {}^2_1D \rightarrow {}^3_2He + {}^1_0n$$

The optimum energy to initiate this reaction is 15 keV, only slightly higher than the optimum for the D-T reaction.

The first branch does not produce neutrons, but it does produce tritium, so that a D-D reactor will not be completely tritium-free, even though it does not require an input of tritium or lithium.

Most of the tritium produced would be burned before leaving the reactor, which would reduce the handling of tritium, but would produce more neutrons, some of which are very energetic.

The neutron from the second branch has an energy of only 2.45 MeV (0.393 pJ), whereas the neutron from the D-T reaction has an energy of 14.1 MeV (2.26 pJ), resulting in a wider range of isotope production and material damage.

Assuming complete tritium burn-up, the reduction in the fraction of fusion energy carried by neutrons would be only about 18%, so that the primary advantage of the D-D fuel cycle is that tritium breeding would not be required.

Other advantages are independence from scarce lithium resources and a somewhat softer neutron spectrum.

The disadvantage of D-D compared to D-T is that the energy confinement time (at a given pressure) must be 30 times longer and the power produced (at a given pressure and volume) would be 68 times less.

p-11B fuel cycle

If aneutronic fusion is the goal, then the most promising candidate may be the Hydrogen-1 (proton)/boron reaction:

$$^1_1H + {}^{11}_5B \rightarrow 3\,{}^4_2He$$

Under reasonable assumptions, side reactions will result in about 0.1% of the fusion power being carried by neutrons.

At 123 keV, the optimum temperature for this reaction is nearly ten times higher than that for the pure hydrogen reactions, the energy confinement must be 500 times better than that required for the D-T reaction, and the power density will be 2500 times lower than for D-T.

Since the confinement properties of conventional approaches to fusion such as the tokamak and laser pellet fusion are marginal, most proposals for aneutronic fusion are based on radically different confinement concepts, such as the Polywell and the Dense Plasma Focus.

The idea of using human-initiated fusion reactions was first made practical for military purposes in nuclear weapons.

In a hydrogen bomb, the energy released by a fission weapon is used to compress and heat fusion fuel, beginning a fusion reaction that releases a large amount of neutrons that increases the rate of fission.

The first fission-fusion-fission-based weapons released some 500 times more energy than early fission weapons.

Attempts at controlling fusion for power production had already started by this point.

Registration of the first patent related to a fusion reactor by the United Kingdom Atomic Energy Authority, the inventors being Sir George Paget Thomson and Moses Blackman, dates back to 1946.

This was the first detailed examination of the pinch concept, and small efforts to experiment with the pinch concept started at several sites in the UK.

Around the same time, an expatriate German Ronald Richter proposed the Huemul Project in Argentina, announcing positive results in 1951.

Although these results turned out to be false, it sparked off intense interest around the world.

The UK pinch programs were greatly expanded, culminating in the ZETA and Sceptre devices.

In the US, pinch experiments like those in the UK started at the Los Alamos National Laboratory. Similar devices were built in the USSR after data on the UK program was passed to them by Klaus Fuchs.

At Princeton University a new approach developed as the stellarator, and the research establishment formed there continues to this day as the Princeton Plasma Physics Laboratory.

Not to be outdone, Lawrence Livermore National Laboratory entered the field with their own variation, the magnetic mirror.

These three groups have remained the primary developers of fusion research in the US to this day.

In the time since these early experiments, two new approaches developed that have since come to dominate fusion research.

The first was the tokamak approach developed in the Soviet Union, which combined features of the stellarator and the pinch to produce a device that dramatically outperformed either.

The majority of magnetic fusion research to this day has followed the tokamak approach. In the late 1960s the concept of "mechanical" fusion through the use of lasers was developed in the US, and Lawrence Livermore switched their attention from mirrors to lasers over time.

Civilian applications are still being developed. Although it took less than ten years for fission to go from military applications to civilian fission energy production, it has been very different in the fusion energy field; more than fifty years have already passed since the first fusion reaction took place and sixty years since the first attempts to produce controlled fusion power, without any commercial fusion energy production plant coming into operation.

A major area of study in early fusion power research is the "pinch" concept.

Pinch is based on the fact the plasmas are electrically conducting.

By running a current through the plasma, a magnetic field will be generated around the plasma.

This field will, according to Lenz's law, create an inward directed force that causes the plasma to collapse inward, raising its density.

Denser plasmas generate denser magnetic fields, increasing the inward force, leading to a chain reaction.

If the conditions are correct, this can lead to the densities and temperatures needed for fusion.

The difficulty is getting the current into the plasma; this is solved by inducing the current in the plasma by induction from an external magnet, which also produces the external field against which the internal field acts.

Pinch was first developed in the UK in the immediate post-war era.

Starting in 1947 small experiments were carried out and plans were laid to build a much larger machine.

When the Huemul results hit the news, James L. Tuck, a UK physicist working at Los Alamos, introduced the pinch concept in the US and produced a series of machines known as the Perhapsatron.

In the Soviet Union, unbeknownst to the west, a series of similar machines were being built.

All of these devices quickly demonstrated a series of instabilities when the pinch was applied, which broke up the plasma column long before it reached the densities and temperatures required for fusion.

In 1953 Tuck and others suggested a number of solutions to these problems.

The largest "classic" pinch device was the ZETA, including all of these upgrades, starting operations in the UK in 1957.

In early 1958 John Cockcroft announced that fusion had been achieved in the ZETA, an announcement that made headlines around the world.

When physicists in the US expressed concerns about the claims they were initially dismissed. However, US experiments demonstrated the same neutrons, although measurements suggested these could not be from fusion reactions.

The neutrons seen in the UK were later demonstrated to be from different versions of the same instability processes that plagued earlier machines.

Cockcroft was forced to retract the fusion claims, and the entire field was tainted for years. ZETA ended its experiments in 1968, and most other pinch experiments ended shortly after.

In 1974 a study of the ZETA results demonstrated an interesting side-effect; after an experimental run ended, the plasma would enter a short period of stability.

This led to the reversed field pinch concept which has seen some level of development since. Recent work on the basic concept started as a result of the appearance of the "wires array" concept in the 1980s, which allowed a more efficient use of this technique.

The Sandia National Laboratory runs a continuing wire-array research program with the Zpinch machine.

In addition, the University of Washington's ZaP Lab has shown quiescent periods of stability hundreds of times longer than expected for plasma in a Z-pinch configuration, giving promise to the confinement technique.

In 1995, the staged Z-pinch concept was introduced by a team of scientists from University of California Irvine (UCI).

This scheme can control one of the most serious instabilities that normally disintegrate a conventional Z-pinch before the final implosion.

The concept is based on a complex load of radiative liner plasma embedded with a target plasma.

During implosion the outer surface of the liner plasma becomes unstable but the target plasma remains remarkably stable, up until the final implosion, generating a very high energy density stable target plasma.

The heating mechanisms are shock heating, adiabatic compression and trapping of charge particles produced in the fusion reaction by a very strong magnetic field, which develops between the liner and the target.

The U.S. fusion program began in 1951 when Lyman Spitzer began work on a stellarator under the code name Project Matterhorn.

His work led to the creation of the Princeton Plasma Physics Laboratory, where magnetically confined plasmas are still studied.

Spitzer planned an aggressive development project of four machines, A, B, C, and D. A and B were small research devices, C would be the prototype of a power-producing machine, and D would be the prototype of a commercial device.

A worked without issue, but even by the time B was being used it was clear the stellarator was also suffering from instabilities and plasma leakage. Progress on C slowed as attempts were made to correct for these problems.

At Lawrence Livermore, the magnetic mirror was the preferred approach.

The mirror consisted of two large magnets arranged so they had strong fields within them, and a weaker, but connected, field between them.

Plasma introduced in the area between the two magnets would "bounce back" from the stronger fields in the middle.

Although the design would leak plasma through the mirrors, the rate of leakage would be low enough that a useful fusion rate could be maintained.

The simplicity of the design was supposed to make up for its lower performance. In practice the mirror also suffered from mysterious leakage problems, and never reached the expected performance.

A new approach was outlined in the theoretical works fulfilled in 1950–1951 by I.E.

Tamm and A.D. Sakharov in the Soviet Union, which first discussed a tokamak-like approach.

Experimental research on those designs began in 1956 at the Kurchatov Institute in Moscow by a group of Soviet scientists led by Lev Artsimovich.

The tokamak essentially combined a low-power pinch device with a low-power simple stellarator.

The key was to combine the fields in such a way that the particles orbited within the reactor a particular number of times, today known as the "safety factor".

The combination of these fields dramatically improved confinement times and densities, resulting in huge improvements over existing devices.

The group constructed the first tokamaks, the most successful being the T-3 and its larger version T-4.

T-4 was tested in 1968 in Novosibirsk, producing the first quasistationary thermonuclear fusion reaction ever.

The tokamak was dramatically more efficient than the other approaches of that era, by a factor of 10 to 100 times.

When they were first announced, the international community was highly skeptical.

However, a British team was invited to see T-3, and after measuring it in depth they released their results that confirmed the Soviet claims.

A burst of activity followed as many planned devices were abandoned and new tokamaks were introduced in their place — the C model stellarator, then under construction after many redesigns, was quickly converted to the Symmetrical Tokamak and the stellarator was abandoned.

Through the 1970s and 80s great strides in understanding the tokamak system were made.

A number of improvements to the design are now part of the "advanced tokamak" concept, which includes non-circular plasmas, internal diverters and limiters, often superconducting magnets, and operate in the so-called "H-mode" island of increased stability.

Two other designs have also become fairly well studied; the compact tokamak is wired with the magnets on the inside of the vacuum chamber, while the spherical tokamak reduces its cross section as much as possible.

The tokamak dominates modern research, where very large devices like ITER are expected to pass several milestones toward commercial power production, including a burning plasma with long burn times, high power output, and online fueling.

There are no guarantees that the project will be successful; previous generations of tokamak machines have uncovered new problems many times.

But the entire field of high temperature plasmas is much better understood now than formerly, and there is considerable optimism that ITER will meet its goals. If successful, ITER would be followed by a "commercial demonstrator" system, similar in purpose to the very earliest power-producing fission reactors built in the era before wide-scale commercial deployment of larger machines started in the 1960s and 1970s.

Even if these goals are met, there are a number of major engineering problems remaining, notably finding suitable "low activity" materials for reactor construction, demonstrating secondary systems including practical tritium extraction, and building reactor designs that allow their reactor core to be removed when its materials becomes embrittled due to the neutron flux.

Practical commercial generators based on the tokamak concept are far in the future.

The public at large has been disappointed, as the initial outlook for practical fusion power plants was much rosier; a pamphlet from the 1970s printed by General Atomic stated that "Several commercial fusion reactors are expected to be online by the year 2000."

The technique of implosion of a microcapsule irradiated by laser beams, the basis of laser inertial confinement, was first suggested in 1962 by scientists at Lawrence Livermore National Laboratory, shortly after the invention of the laser itself in 1960.

Lasers of that era were very low powered, but low-level research using them nevertheless started as early as 1965.

A great advance in the field was John Nuckolls' 1972 paper that ignition would require lasers of about 1 kJ, and efficient burn around 1 MJ. Kilo-J lasers were just beyond the state of the art at the time, and his paper sparked off a tremendous development effort to produce devices of the required power.

Early machines used a variety of approaches to attack one of two problems — some focused on fast delivery of energy, while others were more interested in beam smoothness.

Both were attempts to ensure the energy delivery would be smooth enough to cause a uniform implosion.

However, these experiments demonstrated a serious problem; laser wavelengths in the infrared range of frequencies lost a tremendous amount of energy before compressing the fuel.

Important breakthroughs in laser technology were made at the Laboratory for Laser Energetics at the University of Rochester, where scientists used frequency-tripling crystals to transform the infrared laser beams into ultraviolet beams.

By the late 1970s great strides had been made in laser power, but with each increase new problems were found in the implosion technique that suggested even more power would be required. By the 1980s these increases were so large that using the concept for generating net energy seemed remote.

Most research in this field turned to weapons research, always a fallback line of research, as the implosion concept is somewhat similar to hydrogen bomb operation.

Work on very large versions continued as a result, with the very large National Ignition Facility in the US and Laser Mégajoule in France supporting these research programs.

More recent work has demonstrated that significant savings in the required laser energy are possible using a technique known as "fast ignition".

The savings are so dramatic that the concept appears to be a useful technique for energy production once again, so much so that it is a serious contender for pre-commercial development.

There are proposals to build an experimental facility dedicated to the fast ignition approach, known as HiPER.

At the same time, advances in solid state lasers appear to improve the "driver" systems' efficiency by about ten times (to 10- 20%), savings that make even the large "traditional" machines almost practical, and might allow the fast ignition concept to outpace the magnetic approaches in further development.

The laser-based concept has other advantages.

The reactor core is mostly exposed, as opposed to being wrapped in a huge magnet as in the tokamak.

This makes the problem of removing energy from the system somewhat simpler, and should mean that performing maintenance on a laser-based device would be much easier, such as core replacement.

Additionally, the lack of strong magnetic fields allows for a wider variety of low-activation materials, including carbon fiber, which would reduce both the frequency of such neutron activations and the rate of irradiation to the core.

In other ways, the program has many of the same problems as the tokamak; practical methods of energy removal and tritium recycling need to be demonstrated.

Developing materials for fusion reactors has long been recognized as a problem nearly as difficult and important as that of plasma confinement, but it has received only a fraction of the attention.

The neutron flux in a fusion reactor is expected to be about 100 times that in existing pressurized water reactors (PWR).

Each atom in the blanket of a fusion reactor is expected to be hit by a neutron and displaced about a hundred times before the material is replaced.

Furthermore the high-energy neutrons will produce hydrogen and helium by way of various nuclear reactions that tends to form bubbles at grain boundaries and result in swelling, blistering or embrittlement.

There is also a need for materials whose primary components and impurities do not result in long-lived radioactive wastes.

Finally, the mechanical forces and temperatures are large, and there may be frequent cycling of both.

The problem is exacerbated because realistic material tests must expose samples to neutron fluxes of a similar level for a similar length of time as those expected in a fusion power plant.

Such a neutron source is nearly as complicated and expensive as a fusion reactor itself would be.

Proper materials testing will not be possible in ITER, and a proposed materials testing facility, IFMIF, was still at the design stage as of 2005.

The material of the plasma facing components (PFC) is a special problem.

The PFC do not have to withstand large mechanical loads, so neutron damage is much less of an issue.

They do have to withstand large thermal loads, up to 10 MW/m^2, which is a difficult but solvable problem.

Regardless of the material chosen, the heat flux can only be accommodated without melting if the distance from the front surface to the coolant is not more than a centimeter or two.

The primary issue is the interaction with the plasma. One can choose either a low-Z material, such as graphite or beryllium, or a high-Z material, usually tungsten with molybdenum as a second choice.

Use of liquid metals (lithium, gallium, tin) has also been proposed, e.g., by injection of 1–5 mm thick streams flowing at 10 m/s on solid substrates.

If graphite is used, the gross erosion rates due to physical and chemical sputtering would be many meters per year, so one must rely on redeposition of the sputtered material.

The location of the redeposition will not exactly coincide with the location of the sputtering, so one is still left with erosion rates that may be prohibitive.

An even larger problem is the tritium co-deposited with the redeposited graphite.

The tritium inventory in graphite layers and dust in a reactor could quickly build up to many kilograms, representing a waste of resources and a serious radiological hazard in case of an accident.

The consensus of the fusion community seems to be that graphite, although a very attractive material for fusion experiments, cannot be the primary PFC material in a commercial reactor.

The sputtering rate of tungsten by the plasma fuel ions is orders of magnitude smaller than that of carbon, and tritium is much less incorporated into redeposited tungsten, making this a more attractive choice.

On the other hand, tungsten impurities in a plasma are much more damaging than carbon impurities, and self-sputtering of tungsten can be high, so it will be necessary to ensure that the plasma in contact with the tungsten is not too hot (a few tens of eV rather than hundreds of eV).

Tungsten also has disadvantages in terms of eddy currents and melting in off-normal events, as well as some radiological issues.

In fusion research, achieving a fusion energy gain factor Q = 1 is called breakeven and is considered a significant although somewhat artificial milestone. Ignition refers to an infinite Q, that is, a self-sustaining plasma where the losses are made up for by fusion power without any external input.

In a practical fusion reactor, some external power will always be required for things like current drive, refueling, profile control, and burn control.

A value on the order of Q = 20 will be required if the plant is to deliver much more energy than it uses internally.

Despite many differences between possible designs of power plant, there are several systems that are common to most.

A fusion power plant, like a fission power plant, is customarily divided into the nuclear island and the balance of plant.

The balance of plant converts heat into electricity via steam turbines; it is a conventional design area and in principle similar to any other power station that relies on heat generation, whether fusion, fission or fossil fuel based.

The nuclear island has a plasma chamber with an associated vacuum system, surrounded by plasma-facing components (first wall and divertor) maintaining the vacuum boundary and absorbing the thermal radiation coming from the plasma, surrounded in turn by a blanket where the neutrons are absorbed to breed tritium and heat a working fluid that transfers the power to the balance of plant.

If magnetic confinement is used, a magnet system, using primarily cryogenic superconducting magnets, is needed, and usually systems for heating and refueling the plasma and for driving current.

In inertial confinement, a driver (laser or accelerator) and a focusing system are needed, as well as a means for forming and positioning the pellets.

Although the standard solution for electricity production in fusion power plant designs is conventional steam turbines using the heat deposited by neutrons, there are also designs for direct conversion of the energy of the charged particles into electricity.

These are of little value with a D-T fuel cycle, where 80% of the power is in the neutrons, but are indispensable with aneutronic fusion, where less than 1% is.

Direct conversion has been most commonly proposed for open-ended magnetic configurations like magnetic mirrors or Field-Reversed Configurations, where charged particles are lost along the magnetic field lines, which are then expanded to convert a large fraction of the random energy of the fusion products into directed motion.

The particles are then collected on electrodes at various large electrical potentials. Typically the claimed conversion efficiency is in the range of 80%, but the converter may approach the reactor itself in size and expense.

Inertial confinement fusion implosion on the Nova laser
creates "microsun" conditions of tremendously high
density and temperature

Advantages

Fusion power would provide much more energy for a given weight of fuel than any technology currently in use, and the fuel itself (primarily deuterium) exists abundantly in the Earth's ocean: about 1 in 6500 hydrogen atoms in seawater is deuterium.

Although this may seem a low proportion (about 0.015%), because nuclear fusion reactions are so much more energetic than chemical combustion and seawater is easier to access and more plentiful than fossil fuels, fusion could potentially supply the world's energy needs for millions of years.

Despite being technically non-renewable, fusion power has many of the benefits of renewable energy sources (such as being a long-term energy supply and emitting no greenhouse gases) as well as some of the benefits of the resource-limited energy sources as hydrocarbons and nuclear fission (without reprocessing).

Like these currently dominant energy sources, fusion could provide very high power-generation density and uninterrupted power delivery (due to the fact that it is not dependent on the weather, unlike wind and solar power).

Another aspect of fusion energy is that the cost of production does not suffer from diseconomies of scale.

The cost of water and wind energy, for example, goes up as the optimal locations are developed first, while further generators must be sited in less ideal conditions.

With fusion energy, the production cost will not increase much, even if large numbers of plants are built.

Some problems which are expected to be an issue in this century such as fresh water shortages can alternatively be regarded as problems of energy supply.

For example, in desalination plants, seawater can be purified through distillation or reverse osmosis.

However, these processes are energy intensive.

Even if the first fusion plants are not competitive with alternative sources, fusion could still become competitive if large-scale desalination requires more power than the alternatives are able to provide.

A scenario has been presented of the effect of the commercialization of fusion power on the future of human civilization.

ITER and later Demo are envisioned to bring online the first commercial nuclear fusion energy reactor by 2050.

Using this as the starting point and the history of the uptake of nuclear fission reactors as a guide, the scenario depicts a rapid take up of nuclear fusion energy starting after the middle of this century.

On May 30, 2009, the US Lawrence Livermore National Laboratory (LLNL), announced the creation of a high-energy laser system, the National Ignition Facility, which can heat hydrogen atoms to temperatures only existing in nature in the cores of stars.

The new laser is expected to have the ability to produce, for the first time, more energy from controlled, inertially confined nuclear fusion than was required to initiate the reaction.

On January 28, 2010, the LLNL announced tests using all 192 laser beams, although with lower laser energies, smaller hohlraum targets, and substitutes for the fusion fuel capsules.

More than one megajoule of ultraviolet energy was fired into the hohlraum, beating the previous world record by a factor of more than 30.

The results gave the scientists confidence that they will be able to achieve ignition in more realistic tests scheduled to begin in the summer of 2010.

NIF researchers are currently conducting a series of "tuning" shots to determine the optimal target design and laser parameters for high-energy ignition experiments with fusion fuel in the coming months.

Two firing tests were performed on October 31, 2010 and November 2, 2010.

On March 15, 2012, the NIF's array of 192 lasers fired a shaped pulse of energy that generated 411 trillion watts of peak power - 1,000 times more than whole of the United States uses at any one moment.

The total energy created as the pulse was generated, was calculated to be 2.03 million joules, making the NIF the world's first 2MJ ultraviolet laser – about 100 times more powerful than any other laser in existence.

"Mike Dunne, the National Ignition Facility's director for laser fusion energy, is expecting the giant laser system to generate fusion with energy gain, or "burn", by the end of 2012".

BIBLIOGRAPHY

^ ITER and the Promise of Fusion Energy.

^ "First successful integrated experiment at National Ignition Facility announced". General Physics. PhysOrg.com. October 8, 2010. Retrieved 2010-10-09.

^ "Beyond ITER". The ITER Project. Information Services, Princeton Plasma Physics Laboratory. Archived from the original on 7 November 2006. Retrieved 5 February 2011. - Projected fusion power timeline

^ "Fission and fusion can yield energy".

^ Atzeni, Stefano (2009). The Physics of Inertial Fusion. USA: Oxford Science Publications. pp. 12–13. ISBN 978-0-19-956801-7.

^ a b "Thinkquest: D-T reaction". Retrieved 12 June 2010

^ Iiyoshi, A; H. Momota, O Motojima, et al. (October 1993). "Innovative Energy Production in Fusion Reactors". National Institute for Fusion Science NIFS: 2–3. Retrieved 14 February 2012.

^ "Nuclear Fusion Power, Assessing fusion power".

^ Rolfe, A. C. (1999). "Remote Handling JET Experience". Nuclear Energy 38 (5): 6. ISSN 0140-4067. Retrieved 10 April 2012.

^ Heindler and Kernbichler, Proc. 5th Intl. Conf. on Emerging Nuclear Energy Systems, 1989, pp. 177-82. See also Residual radiation from a p–11B reactor

^ British Patent 817681, available here

^ The first A-bomb shot dates back to July 16, 1945 in Alamogordo (New Mexico desert), while the first civilian fission plant was connected to the electric power network on June 27, 1954 in Obninsk (Russia).

^ The first H-bomb, Ivy Mike, was detonated on Eniwetok, an atoll of the Pacific Ocean, on November 1, 1952 (local time).

^ UW A&A - Research Labs

^ Details of this concept are shown in various publications available on the web page of MIFTI

^ Nathaniel Fisch, "Edward Teller Centennial Symposium", pg 118

^ Great Soviet Encyclopedia, 3rd edition, entry on "Токамак", available online here

^ a b c "The Advent of Clean Nuclear Fusion: Super-performance Space Power and Propulsion", Robert W. Bussard, Ph.D., 57th International Astronautical Congress, October 2–6, 2006

^ "Status of the U. S. program in magneto-inertial fusion", Y. C. Francis Thio Ph.D., Program Manager, U. S. Department of Energy, Office of Fusion Energy Sciences, Germantown, MD, USA

^ "ADVANCES TOWARDS PB11 FUSION WITH THE DENSE PLASMA FOCUS", Eric Lerner, Lawrenceville Plasma Physics, 2008

^ Huizenga, John R. (1993), Cold Fusion: The Scientific Fiasco of the Century, Oxford and New York: Oxford University Press, p. 112, ISBN 0-19-855817-1

^ http://www.discover.com/issues/jan-06/features/physics/

^ Browne 1989, Close 1992, Huizenga 1993, Taubes 1993

^ a b Browne 1989

^ Chang, Kenneth (2004-03-25). "US will give cold fusion a second look". The New York Times. Retrieved 2009-02-08

^ Voss 1999, Platt 1998, Goodstein 1994, Van Noorden 2007, Beaudette 2002, Feder 2005, Hutchinson 2006, Kruglinksi 2006, Adam 2005

^ William J. Broad (31 October 1989). "Despite Scorn, Team in Utah Still Seeks Cold-Fusion Clues". The New York Times: pp. C1.

^ Randy 2009

^ "'Cold fusion' rebirth? New evidence for existence of controversial energy source" (Press release). American Chemical Society.

^ Hagelstein et al. 2004

^ Feder 2005

^ Choi 2005, Feder 2005, US DOE 2004

^ Atzeni, Stefano (2009). The Physics of Inertial Fusion. USA: Oxford Science Publications. pp. 42. ISBN 978-0-19-956801-7

^ Harms, A (2000). Principles of Fusion Energy. USA: World Scientific. pp. 47–56. ISBN 978-981-238-033-3.

^ Pfalzner, Susanne (2006). An Introduction to Inertial Confinement Fusion. USA: Taylor & Francis. pp. 19–20. ISBN 0-7503-0701-3

^ a b T. Hamacher and A.M. Bradshaw (October 2001). "Fusion as a Future Power Source: Recent Achievements and Prospects" (PDF). World Energy Council. Archived from the original on 2004-05-06.

^ Petrangeli, Gianni (2006). Nuclear Safety. Butterworth-Heinemann. p. 430. ISBN 978-0-7506-6723-4.

^ Energy for Future Centuries

^ Dr. Eric Christian, Et al.. "Cosmicopia". NASA. Retrieved 2009-03-20.

^ "The current EU research programme". FP6.

^ "The Sixth Framework Programme in brief".

^ Robert F. Heeter, et al.. "Conventional Fusion FAQ Section 2/11 (Energy) Part 2/5 (Environmental)"

^ Dr. Frank J. Stadermann. "Relative Abundances of Stable Isotopes". Laboratory for Space Sciences, Washington University in St. Louis

^ J. Ongena and G. Van Oost. "Energy for Future Centuries" (PDF). Laboratorium voor Plasmafysica– Laboratoire de Physique des Plasmas Koninklijke Militaire School– Ecole Royale Militaire; Laboratorium voor Natuurkunde, Universiteit Gent. pp. Section III.B. and Table VI

^ EPS Executive Committee. "The importance of European fusion energy research". The European Physical Society

^ Sing Lee and Sor Heoh Saw. "Nuclear Fusion Energy-Mankind's Giant Step Forward".

^ Fusion's False Dawn by Michael Moyer Scientific American March 2010

^ a b "Editorial: Nuclear fusion must be worth the gamble". New Scientist. 7 June 2006

^ physics and engineering basis of multi-functional compact tokamak

^ "US lab debuts super laser", Breitbart news site

^ "Laser fusion test results raise energy hopes". BBC News. January 28, 2010. Retrieved 2010-01-29

^ "Initial NIF experiments meet requirements for fusion ignition". Lawrence Livermore National Laboratory. Retrieved 2010-01-29

^ World's largest laser sets records for neutron yield and laser energy

^ "Most powerful laser pulse in history is fired in US nuclear fusion plant, as scientists bid to harness power of the H-Bomb". Daily Mail. March 22, 2012. Retrieved 2012-03-22

Impulse generator

An Impulse generator is an electrical apparatus which produces very short high-voltage or high-current surges.

Such devices can be classified into two types: impulse voltage generators and impulse current generators.

High impulse voltages are used to test the strength of electric power equipment against lightning and switching surges.

Also, steep-front impulse voltages are sometimes used in nuclear physics experiments.

High impulse currents are needed not only for tests on equipment such as lightning arresters and fuses but also for many other technical applications such as lasers, thermonuclear fusion, and plasma devices.

One form is the Marx generator, after E. Marx who first proposed it in 1923.

This consists of multiple capacitors that are first charged in parallel through charging resistors by a high-voltage, direct-current source and then connected in series and discharged through a test object by a simultaneous spark-over of the spark gaps.

The impulse current generator comprises many capacitors that are also charged in parallel by a high-voltage, low-current, direct-current source, but it is discharged in parallel through resistances, inductances, and a test object by a spark gap.

Made in the USA
Lexington, KY
17 September 2015